End of the World

10 Likely Ways the World May End this Century and How to be Prepared

Table Of Contents

Foreword

Table of Contents

Chapter 1 - Introduction

Chapter 2 - What the Old Manuscripts Say

Chapter 3 - Man-Made Disasters

Chapter 4 - Apocalypse Courtesy of Mother Nature

Chapter 5 - Other Likely Doomsday Scenarios

Chapter 6 - What Really is In Store for Us?

Conclusion

Preview Of 'End of the World: 10 Likely Ways the World May End this Century and How to be Prepared

Foreword

I want to thank you for downloading *End of the World: 10 Likely Ways the World May End this Century and How to be Prepared*

As citizens of our global community, we often wonder about the future of our world. With so many catastrophes, disasters, and diseases seemingly occurring more rapidly, it appears that the world is nearing the end.

This book examines and analyzes 10 likely ways the world could potentially end within this century with research coming from a range of cultural and religious backgrounds. Most importantly, the book will outline how we can be prepared if in case, these doomsday scenarios prove to be true.

Thanks again for downloading this book, I hope you enjoy it!

Chapter 1 – Introduction The Apocalypse, Doomsday, Rapture, the End of Time, the day of Reckoning, Armageddon… Images of vast armies, tortured bodies, mutilated corpses, nuclear infernos, disease and rot! The downfall of human civilization and the end of life as we know it, no matter what you may call it, it all boils down to one gloomy thing! We are all going to die! It may or may not come in this lifetime, but we all know sure as heck, that we will one day be kissing this world goodbye. Together with all our memories and troubles with it! You might ask, what's up with the dark intro? Well, honey, it's better to get that out of the way, don't you think? After all, we just never know, especially now, with all the unstable things going around in our environment, with countries constantly at odds with each other and threatening them with weapons of mass destruction or what not? Not to mention these new found diseases which seem to grow stronger, faster and deadlier and modern medicine just can't keep up with them. And that's only here on Earth! We haven't even talked about the almost limitless number of outside threats from space: Giant, floating debris as big as the entire continent of Asia for starters, then there's this threat of black holes solar flares, and the fact that our good old friend the sun is actually believe it or not, aging rapidly. So, here are the top 10 scenarios on how the world might end, and what to do to in order survive it... who knows, you might even write a better book than this.

FIRST AND FOREMOST, do we know what we are dealing with here?

The answer is both yes and no. But throughout history, there has been many, I mean, countless accounts, reports and written records that go back to pre-Christian years that state that the world is going to end one day. The Bible for instance, and we're talking about both the Old and New Testaments here, From Ezekiel, Jeremiah, and Isaiah, to Christ's apostles. And add to that Book of Revelations (also known as The Apocalypse) by St. John of Patmos, who dedicated a whole section of the Bible about the "Day of Reckoning" itself and more!

The entries and accounts of the prophets of both the Old and New Testaments of the Bible are voluminous to say the least yet, even cultures outside the umbrella of Christianity talks about "The Day." A day of terror for some people, but also the time for salvation as well. The Norse people of Scandinavia for instance, known for the stories of Odin and Thor, and the Nibelungeileid (Ring of Nibelung), hanged a name to Doomsday, they call it Ragnarok, roughly translated as the "Fate of the Gods."

In Islamic eschatology, the Day of Judgment is referred to as Yawm ad-Din. According to their holy Quor'an, this is the final assessment or test by Allah, after which, he will then destroy the whole of humanity, resurrect the righteous and bring forth paradise. Basically, the same as the Christian Apocalypse—which is not strange since both religions are related and tend to share a lot of similarities as well as prophets. Entries pertaining to the end of time also appear in the pages of the Torah, specifically the Tanakh, or the Hebrew Bible. It states that the people of Israel will be released from their bondage and captivity that begun during the Babylonian Era by the appearance of the (Jewish) Messiah, The God of Israel will be known as the one True God. Peace, justice and everlasting life will then follow together with the new Heaven and Earth. The version of the Apocalypse in the Buddhist faith is written in the Pali Canon, specifically "The Sermon of the Seven Suns."

It states that there will soon appear, seven suns in the sky, each having a go at the world, wrecking havoc on the Earth and putting it to ruin. It states that the world would be so hot, that it would appear from a distance as a "single tongue of flame." In Zoroastrianism, the end time is known as Frashokereti, it means "making wonderful." It is the final renovation of the universe and the sole goal of this renovation is for the universe to be in perfect in unity with Ahura Mazda—the god of Zoroastrianism. It involves the total destruction of evil and the realization of paradise.

There are many more "End of days" stories in other religions aside from the popular ones. Because, simply put, if there are stories of the origins of the world/universe, likewise, there would also be stories of its demise. It's just logic.

What does science know so far?

Whenever people talk about Armageddon or Doomsday, supposed learned people will scoff at the idea and take it like a joke. Some would say that it is mainly a spiritual thing, a religious thing, something that was blown out of proportion because of centuries of brainwashing by established religions of the world. They would also note that because of this cultural or religious programming, the concept of the "End Times," will just worsen through the years. It is not pure hokum though, there are plenty of news reports in the 20th century in which many people gave up their lives literally for the sake of religion: There's Jim Jones' People's Temple for instance, where hundreds of people both young and old were made to drink cyanide laced Kool Aid so that they can be saved from the supposed impending doom that was to come. That was in 1976! Fast forward to 1991, twenty years after that, another so called "visionary" appeared in the form of David Koresh, his group, the Branch Davidian told his followers to open fire at police in Waco, Texas! After the firefight, all of the people inside the farm were killed. These two people, wasted and brainwashed their followers to the point where killing their own kind was like their means of showing their allegiance to them. Killing became a means to be saved, to go to heaven and escape judgment here on earth.

Because both Koresh and Jones told them that the end is near, their blind disciples took that as a message that came directly from God Himself.

From the scientific point of view though, researchers and experts both agree that one day the earth will indeed cease to be! There will be no chosen ones or almighty Kings that will rule supreme, there will be no angels versus demons for the souls of mankind and no fire and brimstone. But the scenarios are indeed scary, terrifying and like the prophecies in the scriptures, they all seem to be happening already. The Doomsday scenarios involving science are vast in number as well, on one hand, you have the possibility of giant, continent sized boulders hurling from space and colliding with Earth. On the other hand, an invisible foe, a voracious black hole may be terrorizing the universe eating its way through stars, whole galaxies and planets— which may include ours! Without us knowing it! Some researchers also believe that the end may be a slow dying out process… a whimper, compared to the loud bang today's scientists seem to prefer. And you also have the global pandemic phenomenon involving diseases much like the Bubonic plague which almost wiped out Europe during the so called Dark Ages, in this particular scenario, such diseases will again appear, this time a mutated version, carrying with them antibodies that are resistant to antibiotics and other cures. So devastating is the resulting sickness that it could be weaponized by the military and used for war. Some of these germ weapons were already used in warfare: Most common of them were the Anthrax strain, which when used against animals will result in major starvation, with humans, infectious and deadly, capable of killing huge populations of people. And of course, the most popular and by far the most likely End of Time scenario, World War III! A global conflict so vast and horrifying that the resulting aftermath will be unspeakable to say the least! The terror of nuclear warfare, of the killing, of the millions upon millions of casualties, surviving will be slim at best, and if that wasn't enough, the war's aftermath, Nuclear winter will finish off what remains of the human race. So much death and devastation, that it will take many, many years, before the sky clears out of the pollution caused by the conflicts. Uranium has a half-life of at least a hundred years before it dissipates and becomes inert. If there is still humanity during those times, it will be year 1 all over again, people will have to resort back to living in caves and using rocks for tools. We are going to be hunter-gatherers for the second time around but this time, because of the desolation we've caused, nature will not be on our side. Yabba-dabba-doo indeed!

Chapter 2 – What the Old Manuscripts Say

Documents, scriptures about the end times: Thoughts from Christianity, Judaism, Islam, Buddhism…

The concept of the world ending has been in humankind's history since man learned to tell the days from the nights. When he first noticed how the sun rises in the east and sets in the west, he already recognized a pattern—what comes up, must come down! What begins eventually ends. That is a universal truth, which we are all familiar with. When we talk about the end of the world, or Doomsday, there are always pieces of literature that seem to help us figure it out or at least have an idea about it. After all, stories about the impending Apocalypse, is always a winner! The Bible's version alone is one of the most popular—The Book of Revelations, also known as the Book of Apocalypse, which is still the gold standard as far as great doom and gloom story-telling is concerned (It is not all darkness and damnation I know). It is so detailed, that to this day we are still referring to it whenever some catastrophe has taken place, and the strangest thing about it is we are experiencing some of the events foretold in that book (Some at least...). Which is kind of unnerving in a way but aside from that, there are other cultures that has their own take on that Fateful Day! Some of the most common slash popular ones are here.

A. JUDAISM. They call it "Ha-yamim," The end of days, the Jewish version of the Apocalypse. It has something to do with the appearance of a Jewish Messiah, much like in Christianity which will liberate and gather the exiled diaspora according to their holy book called the Tanakh. Once the messiah appears, His reign will usher the freedom from bondage of the descendants of Moses who originally were the ones with whom the laws of Israel were given to. The books of Isaiah, Jeremiah and Ezekiel contain the basic tenets of the Ha-yamim:

It states these scenarios:

- End of what the world is now.

- The freedom of Israel from captivity, which according to their scriptures, begun with the Babylonian Exile.

- Jews return to Israel.

- The Lord restores the House of David and the Temple of Israel.

- The Lord chooses a regent from the House of David—The Jewish

Messiah, whom will lead the way to an era of justice, peace and love.

- All the nations in the world will recognize that the God of Israel is the one true God.

- The resurrection of the dead.

- The new heaven and new earth. World history will start anew and will begin with a new paradise on earth.

B. ISLAM. The "Yawm al-Qiyamah" or the day of resurrection. It is the final time when God will assess the world, consists of the destruction of everything, followed by resurrection and judgment. Much like the Bible's version of the Apocalypse, the exact time or day in not stated. Also, signs will appear according to the Quoranic Sura, that will give hints that the time of reckoning is at hand. The Sura shows that trouble that would need to come to pass before the end. It speaks of 12 major events that will usher in Al-Quiyamah.

- Terrible corruption and chaos will be the norm.

- The Mahdi will appear and with the help of Isa (Islamic name for Jesus) they will go into war with Masih ad-Dajjal, whom they will defeat.

- The people of Islam will be rid of cruelty. Followed by an era of peace and serenity.

Like some religions, Islam also teaches the resurrection of the dead, a final trial or tribulation and the separation of the good and the wicked. The Muslim version of Armageddon in Islam is known as Fitnah or Malahim. In Shi'ate sect it is known as Ghayba, the righteous are rewarded in Janna, while the vile and wicked will suffer mercilessly in Janannam.

C. BUDDHISM. The Buddha describes the ultimate end of the world in his "Sermon of the Seven Suns" contained in the Pali Canon of Buddhism. It tells the time when the appearance of seven suns will mark the ruination of the earth. Each sun will have its way on desolating the world to bring about the renewal of the universe.

"All things are not permanent; all aspects of existence are unstable and non-eternal. Beings will become so weary and disgusted with the constituent things that they will seek emancipation from them more quickly. There will come a season, O monks, when after hundreds of thousands of years, rains will cease. All seedlings, all vegetation, all plants, grasses and trees will dry up and cease to be...There comes another season after a great lapse of time when a second sun will appear. Now all brooks and ponds will dry up, vanish, cease to be."

—*Anguttara-Nikaya, VII, 6.2 Pali Canon*

Each of the seven suns will cause untold destruction to the world. The words above describe what the first two will do. Then, a third sun will dry out the River Ganges as well as all the rivers on earth. The fourth sun will turn the great lakes into barren wastelands while the fifth will dry out the seas and oceans. The sixth sun will then appear and bake the barren lands until the mountains burn and the skies darken with their smoke. Lastly, the seventh sun will then turn the world into a burning tongue of flame. According to the sermon, all things will burn and exist no more except those who have seen the path.

D. CHRISTIANITY: To many Christians, the central core of the New Testament is the return of the Christ and this Second Coming will usher in the resurrection of the dead and the final judgment. The essential theme for Jesus and his apostles is the last stage of history; the end time is where Jesus will gain His mantle as the sole ruler of earth. In the book of Matthew, Jesus stressed to His disciples the importance of being on the guard for false prophets and false messiahs who will deceive many. He uses the scriptures of the prophet Daniel to warn them about the end of time and the coming of the antichrist. He continues to tell them not to be deceived by their false miracles no matter how incredible they may be. The Lord also says to them that the hour or day of the His coming is not known to anyone, not to angels, not to the Son, but only to the Father.

It is stated in the Book of Revelations, that the antichrist will for a brief period rule over much of the world with his evil powers and influence. Then THE great battle will ensue, possibly the greatest battle ever been waged, the ultimate conflict between good and evil—the war between the real Superpowers, the Armageddon! John of Patmos (also known as John the Revelator), presents in his Book of Revelation, a guide on what to expect about the Apocalypse. The book was originally written for the Christians at Smyrna, Ephesus, Pergamum, Thyatira, Sardis, Philadelphia and Laodicea in order to prepare them for what could possibly be a time of persecution and the return of Jesus.

The Chronicle of Events in the Christian Apocalypse

- The opening of the Seven Seals –
    - A conquering king wearing a crown, and armed with a bow and arrow, the first of the four horsemen of the Apocalypse. Scholars tend to disagree if whether this rider is Christ Himself, ready to do battle with

the forces of evil or the antichrist getting ready to fight Christ.

- A rider on a red horse, the second horseman. The symbol for War, carrying with him a big sword.

- A rider on a black horse carrying a scale. Symbolizes famine.

- The rider of the pale horse. The one that symbolizes death. Natural occurrence of war and famine.

- The fifth seal shows the persecution of Christ's church throughout history but especially during the last days.

- Visions that displayed signs of the great day of wrath— upheavals and chaos, darkening sun, stars falling from the sky, mountains and islands removed and strife the world over.

- The seven trumpets are shown. Heralding the coming of the final and everlasting kingdom.

- Beasts coming out of the abyss. Vast armies of demons, malformed and mutated creatures with forms of scorpions and locusts that bite and sting people to torture them. Also included are demons that look like combinations of humans and animals plus, 200 serpent-like half-horse, half lion men that belch fire, smoke and brimstone. They are all led by Satan himself.

- The antichrist comes to power to deceive the nations. To make matters worse for the followers of Christ. He will pretend to be the messiah himself. John the Revelator was told that this wolf in sheep's clothing can be recognized by a name, and if the letters are made numbers, his name will total to 666.

- Armageddon – The Fateful Battle! According to the Book of Revelation, Christ and his angelic army will defeat The Beast and his dark legions.

- Satan, the Beast and his cohorts are sent to their doom. Satan is bound for a thousand years. With him out of the way, the thousand years of peace begins.

- After the thousand year sentence, Satan is released and goes out to claim back his earthly kingdom. The remaining demons and Gog and Magog becomes his allies once again and sets out for an all out attack against the righteous and followers of God.

- Fire blasts down from heaven. Consuming Satan's evil multitude and Gog and Magog. Satan is held in captive again, this time for eternity.

- The Final judgment—the righteous will be separated from the unclean. The time when God shall the time when God shall judge the secrets of all men and women, and it will be just and thorough.

- The Book of Life is opened. Those whose names are written are given eternal life in happiness and contentment in paradise. For those not included, they will be sentenced to join Satan in the lake of fire—a fire that burns ceaselessly for all time.

Survival in the End Time: Is it really possible?

We're talking about a real cataclysmic event here, one that can really shake the foundations of the earth. In this section, we have 10 scenarios—All life-ending scenarios! Now, it is possible that you might come across one of these situations (We sure hope not…) in this lifetime or the next. Most of the scenarios here are survivable… but in reality, there is really no surviving when we are up against something like a black hole or a gamma ray burst heading straight for earth. What it means is whether you live or die; only you can answer that! The end time scenarios included here are the ones most likely to happen, and as much as possible, we excluded the ones included in the Book of Revelations, or any other scriptures. Because if God wants to end the world tomorrow or today—there is really nothing much we can do about that!

Chapter 3: Man-Made Disasters

1. WORLD WAR III – THE MOTHER OF ALL WARS! THE WAR FORETOLD?

Once the ball gets rolling, there's just no way of stopping it! (unless Divine Intervention takes part…) Once the proverbial Doomsday Clock strikes at midnight, what happens next, is so gruesome, you will wish you should have been killed by the first bullet fired.

There are many possibilities on how this might happen or might begin… The East-West conflict might begin a new; China might wage war on Japan this time or is it North Korea? Libya, Iraq, Lebanon and Syria might join forces into one confederation set its targets against the USA. Al Qaeda might finally get hold for itself a thermonuclear device or get enough funding to buy one in the Black Market, somewhere, or somebody in the USA might simply just press the wrong button—Whatever the reason, doesn't matter anymore! The inconceivable already happened… There's nuclear war! Armageddon has begun. No getting out of it now! Soon, you will see ICBMs crossing the skies, like fiery, featherless birds of destruction, and then… Boom! Blinding light, Inconceivable heat, you're body simply just melts away, your surroundings burn, everything burns… You are dying inside ground zero, inside a thermonuclear inferno, where nothing escapes.

Around you, it's no different. Blinding flashes, ear-splitting bangs as bombs go off left and right. Like a scene in Terminator 2: Judgment Day! The human race is being destroyed by the very weapons meant to protect it. Everywhere you look, giant, mushroom-shaped clouds tower on the horizon. Whole cities are decimated. Infrastructures crumble and turn into smoldering rubble, and of course, the dead... Millions upon millions die! (and that's the mildest estimate) Entire continents become barren, uninhabitable wastelands in weeks. The survival rate dwindles exponentially as more nukes go boom.

But let's say, you were able to survive all that. You were able to toughen up and go on living... For now. Better look for adequate shelter though, for the next days are not going to be much better. In fact, you would probably wish that you were one of the lucky ones who died immediately after the first nuclear strike. It will surely rain nuclear fallout for the next few days or so. It will be followed by doses of radioactivity in several regions, hundreds of kilometers from every impact site. It would be best to just stay indoors at least for the next couple of years or so until all the radioactivity levels drop.

But that's not even half of it, once the nuclear cloud clears up, all the ejected material from the earth including the dust, debris, soot and radioactive particles rise up to the skies and will cover it. The stratosphere will become so blanketed with these gunk that it will cover the sun for days, weeks perhaps months. Temperatures drop drastically paving the way for the dread Nuclear Winter. Within weeks, it is minus 20 or 30 degrees Celsius everywhere, except for coastal areas because the oceans warm up the land. They will drop just minus 5 or 10 degrees Celsius. The downside of this is because of the temperature differences of the inland and the seaside areas, this is an ingredient for huge storms and hurricanes that will terrorize the coastal areas and other low-lying places. Also, the dust and fallout particles that were sent up to the atmosphere will of course eventually come back down to earth in the form of soot and granular ash, bad news when it rains as well because this turns the rain into acid rain, and not the kind you experience in normal suburban condition levels, oh, no! We're talking about real acid… The sulfuric kind burns your skin upon contact. When it's not raining, the wind will blow dust unto your face, not enough to kill you, but will not do you any good in the long run either. Weeks, months pass by, plants, and then animals die. Soon, there will be nothing left to eat. But you are not alone, because not all animals died out, the ones that are left are the ones you would consider to be bad company: Roaches, rats and flies... Enemies! And you're competing with them for everything. And they spread diseases too! Like how we used to know them. Oh, they will have a grand time feasting on the remains of countless dead.

This seems to be getting worse and worse, and we're not done yet! There are still other things: after the breakout of diseases caused by the millions of casualties of the war plus the vermin that is spreading them, outside of your bunker is still no-man's land! The heavy amount of pollution caused by the conflict has obliterated what was once known as the ozone layer. Now, there is nothing to shield us from the harmful solar radiation and ultraviolet heat from the sun. From that point on, every time the sun shines, you need to run to the shadows!

It definitely looks grim. Even if you manage to go through all that and survive, the world as you know it, is gone. With the breakdown of society, the food chain in shambles and humanity dehumanized, we are at the very least back to the stone age, It's year one all over again.

HOW TO SURVIVE NUCLEAR WAR, or at least have a chance to...

Scientists believe that for Nuclear Winter to take effect, a full scale Nuclear War must first happen. If that does come to pass, the effects will be downright terrifying! It would destroy the Earth's atmosphere and hurl dust, soot and debris that could block sunlight for months or even years… ushering the feared Nuclear Winter. The second Ice Age.

A. SURVIVE THE ATTACK

If you are far away from the place where the bomb or missile went off, stay away from windows and do not look at the flash! The shredded glass from the window and the force alone will rip you apart, if not leave you blinded for life.

B. LOCATION, LOCATION, LOCATION

If you live in a great city of a powerful country, like New York for instance, remember that you are in the hit list of candidates for a Nuclear Attack. Manhattan is one of the busiest spots in New York and it is also one of the richest, close to 2 million people either living there or working—A megacity like that is always on the cross hairs of people with an evil agenda.

C. AFTER THE INITIAL STRIKE, WHAT NOW?

So you have survived the bomb. Now here are other things to think about:

- FACT 1: The Nuclear War depleted much of the Ozone Layer. So those harmful ultraviolet radiation can now easily slip through our atmosphere and bombard us with harmful rays

- FACT 2: The dust and debris caused by the nuclear explosions have blanketed the sun. The temperature on earth is now 20 degrees below zero.

- FACT 3: Most plants will die out, then after that, animals will! Global starvation at its worst.

D. BUILD YOURSELF A BUNKER

Clearing out your basement is definitely a good idea, because if things go sideways and you got yourself in a Nuclear Emergency, it is best to have somewhere where you and your family can hide and shelter yourself from harm. If you have a swimming pool, just drain the water out and build a ceiling on top of it. Add 3 feet of dirt to cover the ceiling, and you got yourself a bunker. You need to have at least 18 inches of walls and ceiling.

E. COVER UP

When going outside, be sure that you are covered from head to toe from ultraviolet radiation. Wear a hat when outside, and always stay in the shade. Protect your eyes as well, shades and sunglasses can help you. Use sunscreen lotion for added protection from sunlight, make sure it has a high UV protection factor. Always be mindful of your kids, for they are more susceptible to UV radiation.

F. FUEL, FIREWOOD AND OIL

If a long lasting cold spell is to occur, make sure you have emergency heating supplies. Electric lights and stoves as you should expect will be a luxury. It is best to store lots of firewood around, some matches and an old stove. It is also good if you have some fuel that you can use.

G. MAKE SURE YOU ARE PROPERLY INSULATED

Since you will be living in extreme cold for a while, get used to the dress code! Being properly clothed is a matter of life and death especially in a super harsh climate that is Nuclear Winter. Several layers of clothing will help you maintain your core body temperature and also allow you some movement. Other than that, it will allow the perspiration of your body to disperse more readily. Boots, goggles and gloves should also come in handy.

H. INSULATE YOUR HOUSE

Familiarize yourself with the methods used by people living in the North when it comes to getting their houses winter-ready! Have tools at the ready. Make sure the attic and basements are well insulated. When the temperature gets even worse, add more insulation to key areas of your house to keep you and your family warm.

I. FOOD PLAN AND FOOD RATION
Remember that growing food at this time, is an impossible feat. Ice covers everything, and it's going to be like that for quite a time. Food will definitely be an issue! Your ability to look for food will be a vital factor in your continued survival and it will be an ongoing battle. Initially though, it will be good to have some non perishable food stuff and water ready. A year's worth of supplies might stretch you for quite some time, and with dwindling supplies, you will be responsible in rationing the food. The ability to stretch your resources thin might be your best bet in keeping yourself alive.

J. LASTLY, BE CAREFUL!
Being a survivor means that you are ready to face whatever challenges life has to give you. Remember that social chaos is now the norm. People are starving everywhere and there are plenty of desperate people out there! Theft, robbery, looting and murder might seem to be a logical choice for some people to survive. So always be ready to defend yourself.

Also, this is all hypothetical, it is up to you to assess what is going on around you. Today, if you think that we are in the verge of a Nuclear War, let's get ready!

Movie Reference for Nuclear War Scenarios:
a. World War III (1982)
b. The Day After (1983)
c. Dr. Strangelove (1964)
d. Threads (1984)

2. The Ice Age once again!

With all these talk about Global Warming, it is easy to think that the Earth is warming up. The thing is, the world plunged in another Ice Age, one of the reasons of the dinosaur's disappearance is a much more likely scenario: Imagine, the whole of Europe and the US, covered in ice and permafrost! One more thing, it can happen sooner than we think! The truth is we are already living in an Ice Age.

The Ice Age, like the one we're living now, started 1.6 million years ago, dubbed as The Pleistocene epoch. In fact, one of the best proofs of that Ice Age is still with us: The ice in the polar regions and some mountain tops are still with us right? The reason why most of the climate that we have today in this world is warm is because the Ice Age version that we have right now is what you call Interglacial—a period of milder temperatures in between frost-covered glacial, where the ice just stays on the polar ice caps and mountains. This interglacial period started 12,000 years ago, so the next glacial period may arrive at anytime. The thing about this process, is that it starts rather sudden.

One day, you might just hear on the news that something mind-boggling has happened. Like snow and hale falling in China and some other temperate regions. This event might set the ball rolling for a big glacial event—the process where the ice comes back down from the slopes and polar ice regions. The snow then, acts like a refractive mirror that throws back the heat and light from the sun, hence the earth does not warm itself back up. The glacial era is back online!

Once the climate shift rolls in, cataclysmic weather disturbances will appear. Huge storms and violent earthquakes will follow, not to mention the intense flooding that will happen that will put the world in an ecological disaster of biblical proportions. Millions if not billions will die and the population of this world will be decimated. Of course, the rest of humanity will have to deal with the freeze-kill that's going to follow. Rivers and oceans will be frozen and what used to be farm lands will now be permafrosted tundras. Again, ice will kill the life out of this planet.

Surviving the Ice Age

It's time to prepare for the coming Ice Age. With all the nasty weather developments happening, it is best to be ready before the frost bites you from the behind. They say that every 90,000 to 100,000 years there have been Ice Ages, our last one was 12,000 years ago, we are overdue by 2000 years. Here are the important steps to remember:

1. FIND SHELTER

This is essential, since with the harsh condition brought about by an Ice Age, looking for a place of refuge will be difficult. Look for something that will provide protection from the cold and is strong enough to withstand earthquakes brought about by the Ice Age. Preferrably underground.

2. LEARN TO HUNT.

Being squeamish will get you nowhere. You have to eat meat sometime. And with the protein that you require, you will need quite a lot. Since the fats from the meat can warm you up, you will also need protein, meat is rich with those. Learn to fish as well.

> (The rest of the steps about Ice Age is the same as the steps for Nuclear Winter and Meteor)

Movie reference for Ice Age scenario:
The Day After Tomorrow (2004)

3. Global Pandemic: Diseases Galore

We tend to worry a lot about Global Warming, Asteroid collisions and Alien Abductions and what have you—Yet, the biggest threat to our existence is already here, and doing fine. Diseases! It's one of mankind's oldest nemesis and it wants pumping up for another massive attack, and you know, it could probably wipe us clean off from the face of the planet.

People and germs, have always coexisted. Since the beginning of time, we lived side by side with sickness and disease for much of our existence here on Earth. Sometimes though, the diseases strike with so much fury and poison that it creates panic on a worldwide scale. The diseases such as The Bubonic plague, AIDS, SARS and Ebola will forever be etched in our collective consciousness because of how hard and vicious they kill.

If you can recall, the Bubonic Plague killed half a quarter of the population in Europe during its heyday, the Spanish flu quadrupled the death toll of World War I in 1918. In the US alone, this flu epidemic killed more people than World War II, Korean War and the Vietnam War combined.

Our doctors may be giving the A-OK today, that everything is in control and we are going to be alright. The thing is, to many people—this is a blooming lie! Because it just seems that we are more vulnerable and prone to any kind of deadly bug out there! And it's a real cause for alarm. Since the world is so densely populated, and agriculture does leave us open to deadly pathogens (take for example, the Avian Flu or Mad Cow disease), and also humans love to travel, sometimes they do not care of what they are carrying with them. We've been battling this ongoing war with diseases since time immemorial, haven't we won already? How about the new developments in science that we are all so proud about? Well, the thing is, diseases, like all living things tend to evolve or sometimes mutate.

That is why there will always be new diseases because, evolution itself throws new lethal developments in the mix as well! Sometimes an almost harmless bacteria or fungi crosses the species barrier and is now attacking livestock or humans. The AIDS virus for one, started out like that! From what used to be a disease strictly for monkeys became one of the diseases that continue to be a scourge to humans. Ebola first infected bats in Africa, now it's a plague that ravages the continent.

More bad news for us—because there are a lot of viruses like now, that seem to just come from out of nowhere. At the same time, doctors are baffled on how to treat them because of the limited information about them. What's more scary is that they seem to be resistant to whatever treatment we can come up with. Not to mention, how savage their manner of killing is. They seem to pop up from nowhere and kill whoever they infect with no means of stopping them.

The influenza for one, is probably the oldest of the diseases known to man. The flu virus has been ravaging mankind since recorded history and seems to stay in active circulation. We waged wars with it countless of times, and the best that we came up with was to neutralize it. In its Spanish Influenza form, it was a deadly, wasting disease that left its victim dying in agony. Uncontrollable fever, seizures, blood and foam coming out of the mouth and they die because of drowning, Fluid has filled their lungs. One witness in 1918, stated: "This was not influenza, this was a plague. The world is coming to an end." One cannot possibly underestimate this flu, it kills 90 percent of its victims! It was highly contagious and during that time, incurable. It also evolved so fast that it was almost impossible to find a cure for it.

The Spanish influenza was really nasty, but it may only be a matter of time before we really come across a flu-like strain that will really decimate the global population. SARS, Avian Flu, Anthrax and even Ebola could one day prove to be just a weakling of a strain, compared to the next super flu. As a matter of fact, the flu strain of 2003 has some virologists really worried.

Aside from viruses, there is this ongoing concern about Superbacteria. Can you imagine a time when the most dreaded diseases make a comeback: Measles, smallpox, tuberculosis and the terrifying menace, the Bubonic Plague, and once again blaze their trail of death all over the world. A lot of these strains would soon find their way to us again… but this time, more durable than ever, resistant to antibiotics and other medications. These superstrains already claimed lives in both Japan and Africa. Same thing for the newest mutated strain of Tuberculosis, it is so invulnerable, it is almost impossible to treat.

The thing about these diseases, is that it's like Russian Roulette all the time: Whenever scientists and doctors think they finally have the disease figured out, it suddenly does something unexpected—then boom! You realize that somehow you're fighting a losing fight. It's just that you never know what's going to happen next with these kinds of diseases. There was an incident that happened inside a lab in Australia in 2002, a group of scientists accidentally turned what seemed to be a harmless virus into a very destructive and fearsome killer. Luckily for them, the virus was too unstable to last very long and died shortly after. Think of what would have happened if some bystander or an ordinary pedestrian caught whiff of the virus, think of what chaos that would have bring. Here is another scary thought, what if some mad dictator discovered this and decided to buy the virus and weaponize it or if he wants to discover what is going to happen if he tries to tweak an anthrax strain or tells his scientists to engineer an airborne Ebola virus. We are already in the 21st century, and the last century showed us the awesome power of these microscopic monsters and their killing efficiency. We hope and pray that a global pandemic is not brewing up somewhere in the horizon.

Surviving a global pandemic

A global pandemic may very well turn the world economy to a standstill. Without the adequate manpower to operate machinery, huge mega-structures would be nothing, leaving you without groceries and utilities such as power, water and other services.

If a deadly infection entered the population, and was able to spread uncontrollably from person to person and became a global epidemic, we got a pandemic on our hands!

Survival of a global catastrophe such as a pandemic is focused mainly on how you isolate yourself from other people who may or may not be infected by the contagion. You will have to separate yourself from the rest of the herd so to speak. This is going to go on for several days, months or until the disease has finally settled down or died out.

    A. The Rule of Thumb is, don't wander around especially outside your group.

Make sure you have everything that you need at home, together with the people who are closest to you. It cannot be stressed how important it is to have the ability to self quarantine. Ideally, the best place to do that is inside your house, do not go out unless you really need to, and if you do, always go outside protected. Since, nobody knows for sure who the infected ones are.

B. Know some basic information about the disease.

In the Art of War by the Chinese strategist Sun Tzu, it is said that the first rule of war is to know your enemy (That's not an exact quote), and an essential weapon that you can use during this time of crisis is information. It will save your life! Know the symptoms of the disease, the incubation period especially, because there are some diseases that won't show any symptoms but are already contagious, and there are some that will be contagious once the symptoms begin to show. This is an important piece of information because this will allow you to stay healthy and knowing what to do next.

C. If any one of your group members becomes infected, he or she needs to be quarantined… for the sake of the uninfected ones.

That is also the same procedure when accepting a new member of the group. They should know that themselves and should have the initiative to self quarantine for a number of days depending on the kind of epidemic: Most viral strains show up 3 to 5 days, sometimes 10 to 15, the incubation of Ebola is 2-21 days sometimes even longer.

D. The quarantine area.

Most of the time, the room for quarantine is a separate building—an outhouse, garage, barn and the like, if you plan to offer a specific room inside your own house, it should be vented outside. Negative pressure is also important so leave one of the windows cracked, so the air flows from the remainder of the house and out of the window.

E. Your pandemic survival kit.

These items are essential it you need to get for yourself and your group mates a survival kit:

- Water

- Food

- Bleach for disinfecting
- Exam gloves
- Antibacterial soap
- Toilet roll
- Duct tape
- N95 masks

Movie references for global pandemic scenarios:
Black Death (2010)
The Crazies (2010)
Contagion (2011)
Outbreak (1995)

Chapter 4: Apocalypse Courtesy of Mother Nature

4. Giant Meteor

This is by far the most common and most feared extinction level event that we could think of! Think about it, the information is all over. You can watch it on the tube, there are countless movies out there about this very same event, research on it in the internet and you got yourself an entire library and tons of information about this world ending disaster of mega proportions. According to scientists, it already happened once with the dinosaurs and other prehistoric life… Why shouldn't it happen today? Whether you refer it as The Big One, Lucifer's Hammer, The Great Exterminator or the Global Killer… What it basically is is a giant mountain from space and it is coming down to destroy us!

It just starts out as a big fiery mass in the sky, a few seconds of air time and then: Boom!

The atmosphere is suddenly on fire, a gigantic column of fiery debris will shoot up from the sky from the impact. Hundreds of thousands will die immediately, the ones nearby and everything within a hundred or thousands of miles will be burned to a crisp, the people on ground zero will simply disappear, vaporized. The shockwaves generated will be sent around the globe. Much like a stones makes ripples when it is thrown into a pond, this rippling effect will rearrange everything on the earth's crust! Everything will be decimated. Huge earthquakes, followed by gigantic tsunamis, the likes of which never been seen in human history. Volcanic activity everywhere will be reactivated, millions upon millions die as cities are are tossed, turned and shaken like a ragdoll. On the opposite sides of the planet, both blast waves will slam with each other and create a mountain ridge within seconds. The earth will get a very destructive make-over to say the least.

And that's just the initial phase, oh yes, that's just for starters! As the quakes, the waves and fires of destruction dies down, the huge cloud of debris that got tossed up from the impact sight will then spread out and blanket the stratosphere with carbon, dust and there it will stay for months if not years, temperatures everywhere will drop drastically, some twenty to forty degrees. This chilling effect will soon follow, freezing the earth. Plant life will die within weeks, animals that eat plants will also die of starvation—bad news for us, because we rely on both plants and animals for food.

The few remaining people still hanging on, probably living in mine shaft caves and old bunkers, will have a heartbreaking revelation once they come out of their reclusive shelters, seeing the world as a lifeless, barren and desolate wasteland. Filled with dust and ruins, the human race will just have to start again from scratch.

SURVIVING THE CATACLYSM
A. Listen to the News!

When the NASA tells the whole world the inconceivable truth, We better heed that warning. Usually, once the comet or asteroid gets into Earth's orbit and within striking distance, leaders of the world will have to tell their people the truth regarding our options to avert disaster or if not survive. They will surely give us several months' notice before the time of impact arrives to help humankind be ready for this catastrophic event and of course to plan ahead on what to do regarding the giant projectile heading our way (do they try to destroy it or just brace for impact?).

B. A year to six months before impact... What to do?

Find out where it is going to hit or what the experts say regarding its falling trajectory. Is it going to land on the ocean? On is it going to fall on the ground?

C. Stock up supplies and build your bunkers.

Make your own house an underground fortress to protect you and the ones you love dearly. Find that right spot where you can hide during or after the impact. It should be underground, with no windows that can harm you and strong enough to protect. If your underground area has windows, make sure to brick them up. As for supplies, make sure you have enough food and water to last you for at least a year, do not forget weapons as well, because you never know, once it gets down to it, people will fight to grab whatever means they need to survive. Once, the meteor collides with Earth, our darkest fears will then be realized, so be prepared to fight for what you have and protect the ones you love.

D. Read up! Collect all available knowledge as you can.

Learn everything as much as you can! Survival skills, combat tactics, hunting, farming, growing animals and what have you... You are going to need it. You will be left in a world that has close to nothing, a world that you will have to build from the ground up. Once you surface, all vegetation would have been gone, and no animals left as well. You will have to trust your own instincts and will power to survive and battle the elements. The know-how that you have acquired could be the difference between you surviving this catastrophe or not. Being learned in the military sciences can also keep you alive by not being subjugated by anyone who has skills like yours. Be ready to write them if needed be, so others can also learn and benefit from it.

E. Three months before impact. Is the bunker ready?

If you live in an place nearby the expected area where the object is going to hit (in case for some reason, you cannot leave this area), try to find old bunkers and bomb shelters that were used in the 60s when the cold war was at its highest. These bunkers can withstand nuclear blasts and are your best bet. If you do not have that option better fortify your shelter more... and fast!

F. One month to go. Stock it up nice and high!

Get more supplies if needed be, the more you have, the higher your chances of survival are! Aside from food and water, make sure you have medicine as well, clothes will be a very vital thing, especially thick clothes, since, when the asteroid collides with earth, the aftermath will be like that of a nuclear war, with all the debris, dust and soot that will be thrown up in the air, it will cover sunlight for months perhaps even years, a second ice age could be very, very well be possible scenario. Include seeds also, because once it's alright to go topside, you can start growing food. Also, include mementos and prized possessions, because once you venture topside after the debris has settled, the world will be in a very, very different state. Include toys and various pastimes because it could get really boring underground.

G. Stay indoors!

A few days before impact, close the doors and don't leave your shelter! Make sure you have everything you need.

H. Impact!! Brace yourselves!

Once the rock hits the earth, huge earthquakes will follow, so are debris and dust from the resulting explosion. Dormant volcanoes may suddenly become active again because of the quakes generated by the impact. If you live near a volcano, you should have evacuated a few months beforehand. Tsunamis could be triggered by the meteor impact as well.

I. Ejecta

Ejecta are rocks, dust and debris that rose upon the impact with the meteor. A few hours after the collision, these giant, burning debris will come crashing down back to earth and it will heat up the atmosphere at an average of 100 degrees Celsius. This is the reason why your best shelter is a nuclear fallout bunker.

J. Record your experiences in a diary.

Since it is dark outside because of all the debris thrown to the sky, the sun will not be able to shine through. Just stay in your shelter and write down your experiences, you have a heluva story to tell.

K. It's getting nippy now.

A month after the impact, global temperatures have plummeted, venture out only to gather firewood so you can stay warm.

L. Be careful of bandits.

You're running out of supplies and it is too cold. Venture out of the shelter and go near the ocean, since it is warmer there. Just be sure to carry some heat with you as protection from desperate individuals.

Lastly, just be careful. More or less, it would take about a couple of years before the sun actually shows up again. Once it does, make use of the seeds that you saved and plant them. Hopefully, they will grow and with them, you can extend your survival even more.

Movie References for asteroid impact scenarios:
a. Deep Impact (1998)
b. Armageddon (1998)
c. Asteroid (1997)
d. Meteor   (1979)

5. Gamma Ray Burst – In seconds, we're burning!

They say it is the most destructive killer in the Universe. That we are lucky that it has never crossed hairs with us. If it does strike, well, then it's goodbye to our planet—Because we surely do not stand a chance against this killer from space—the most feared Gamma Ray Burst or GRB.

It's like a lottery, somewhere in the Universe, Everyday, one moon, star or planet gets picked up, and then—Dies in a fiery explosion caused by energy so large, that it's equivalent to the power our sun could generate its whole lifetime. You just do not want to be anywhere near this very power blast. Only recently scientists and researchers learn what a GRB is really is. It is the amount of energy released by a dying star that is being eaten from the inside by a black hole. In those last moments of that star's life, it emits a huge blast of energy almost at light speed—some scientists call it "The Birth Cries of Black Holes."

One problem is, in our part of the Galaxy, the Milky Way, one of them is about to go kaboom! Though this star, Eta Carinae, if it bursts, won't harm us directly since its axis is away from us but, in any given day, there is a possibility that out portion of the galaxy might just get picked for a blasting. If a Gamma Ray Blast does happen, it's going to be equivalent to Hiroshima A-bombs going off to every part of the world—all at the same time. This will blast will obliterate all of the Earth.

SURVIVING A GRB

Surviving this particular disaster is close to impossible, unless of course we can find another suitable planet where we can live in. As of writing this, there is no such planet yet.

Chapter 5: Other Likely Doomsday Scenarios

6. Supervolcano

You think a meteor colliding with Earth is devastating enough? Well, before the dinosaurs even came to being, a volcano did a more thorough cleaning.

During the Permian-Triassic Era, something called a Permian extinction happened. It was caused by a massive volcano eruption. 90-95 percent of those living during that time disappeared. What a lot of scientists are considering is a giant volcano, somewhere what is now known as Siberia, erupted so violently and massive that it reset the Earth's climate and evolution itself.

At this moment, one particular dormant volcano in Yellowstone National Park is being monitored by scientists. This volcano is said to erupt about every 600,000 years. The last time it erupted was 640,000 years ago—so we are long overdue for a cataclysmic event!

7. Machine Uprising

When machines suddenly become self aware, we are dead!

8. Black Hole: An Invisible Enemy

The thing about black holes is that they are an invisible threat, nobody can see them. Even our greatest scientists and researchers are not sure where they can be found. There is no space telescope available that could actually show you where or how it looks light. Since the black holes gravity is so strong, that even light cannot escape it. You can only get the idea that a black hole is around, if it is already about to devour you and whatever planet you were in at the moment.

It could happen any moment and that's what is scary. At first, researchers and scientists could one day view our galaxy and find out that the outer planets of our Solar System's orbits are acting strange, and then some planets actually change course. When it finally dawns to them that the reason for this is a black hole, the earth may only have a few months to live, perhaps a year before it finally gets to us and swallow us in. Once, we are at striking distance of a black hole, well, there is really nothing else we could do, the planet will either be wrenched out of our orbit and zoom straight to the hole or it could mess with our gravity, catapulting us out of orbit and throwing us out to where the outer planets where.

So, our end will either be being sucked inside the black hole, into nothingness. No woes of agony or explosions for those will be swallowed by nothingness as well or our planet might become a lost planet, out of orbit and far away from the sun, covered in darkness and cold forever.

Surviving a black hole

The possibility of surviving a black hole kind of Apocalypse is ludicrous to say the least. Since, once the gravity of the black hole devours gets close enough, within seconds, you will be sucked into it… So there is really no way for you to survive this. The only thing that can be done is maximize all the time that you have left. Movie reference for black hole scenario: Supercollider (2013)

9. Zombie Apocalypse (Rrruaarrr…)

As of writing, this is probably the most famous of all the apocalyptical scenarios here. It is probably because of the resurgence of the zombie genre in the movies.

It all starts one day, you're minding your own business when suddenly people around you, start to jump at you and bury their teeth into you— Blood gushes out, you faint! Seconds later, you wake up (in a way), finding out that your personality and humanity is gone, you're half alive, part decaying flesh. Unable to think or speak, you can only feel that you are hungry… Extremely hungry.

The zombie scenario comes in many forms though: One version involves a virus mutating us into mindless, blood-thirsty, living dead. One other scenario, a chemical mishap suddenly wakes up the dead. It remains unclear, and unrealistic to some—How can a corpse who is almost completely decayed suddenly, able to stand and sometimes run around hunting for food. What are they hungry about anyway? They're dead!

One possible scenario today, especially after the outbreak of diseases like Ebola, is the zombie apocalypse might come from viruses that actually can mess up the brain, like in the movie 28 weeks Later (In that movie, it was a Rabies epidemic), if the world does end like that, living in the modern world will definitely be a pain in the neck.

HOW TO SURVIVE A ZOMBIE APOCALYPSE

1. Research everything about zombies, because obviously the internet is still working.

2. Shop and stock up your supplies and I mean all of them. A year's supply will be enough for now—but of course more is always welcome.

    a. Medicine

b. Food

  c. Water

  d. Weapons

  e. Toiletries

3. Setup even the most basic communications center.

4. Diesel is best.

5. Practice your marksmanship.

6. Learn to fish and hunt.

7. Find ways of making electricity

   These are all practical ways to surviving any calamity, and not only a zombie initiated one.

   Movie Reference for Zombie scenario:
   a. Night of the Living Dead (1968)

   b. 28 Days Later (2002)

   c. Grindhouse – Evil Dead (2007)

   d. World War Z – (2013)

   e. Resident Evil films (2002-2012)

f. The Walking Dead Series – (2010 onwards)

10. Alien Invasion

The aliens that might visit this planet may not be the nice ones we were hoping for.

Chapter 6 – What Really is In Store for Us?

To answer this question, mankind needs to look inside himself and ponder, why are we here on earth in the first place? What is so special about us humans? The only difference that we have really with other animals is our highly developed thought process—our powerful minds! We don't have wings, nor fangs, prehensile tails or brute strength, yet we tower among the animals as the major consumer of resources. We are top of the food chain so to speak, because we can impact the natural flow of things directly. We can initiate change, we can build and of course we can destroy. To some extent, we liken ourselves to gods (at least of the beasts and plants). But still, the question is there, why are we here? Do we have a great impact on the grand scheme of things? In terms of the magnificence of the universe, we are just a tiny, miniscule speck. A tiny pimple in the blanket of space! The great book tells us that the Almighty created us in His own image, that we were destined for great things since day one. That, and the authority God gave us to lord over the seas, lands and air. But now, it is somewhat getting apparent, that our glorious dreams of travelling space, conquering planets, teleporting from one place to another or just driving gravity-defying cars might remain just that—a dream! A figment of our imagination, a delusion! No future where our skills and learning can just be downloaded via internet ala The Matrix films, where we can just hit the light speed lever in a train like in Star Wars, and be in the North Pole then back again in a blink of an eye. We could go on and on! The thing is we just don't know anymore what the future holds or if there is a future to speak of. However, on the other hand, there might still be a future! With all the marvels science has achieved over the years! The treasure trove that is the information superhighway! The endless potential for curing many diseases (now that the human genome has been effectively mapped) and improving the quality of life! Yes, doomsayer or doom seers had always been talking, moaning and screaming like the sky was falling since recorded time! In the year 1000AD, for instance, many pilgrims in Europe sold everything they had so they can go to Jerusalem and wait for the return of the Christ. A monk from Yorkshire, England by the name of Druthmar even gave an actual date, 24th of March, much to the dismay of the people. Let's put it this way—a lot of people got a little nasty after that date. After they realized that they have all been duped and taken advantage of. Strangely, many people are still being fooled by this ruse, especially after the dawn of the telecommunication marvels of telephone, radio, television and of course, the most powerful tool in modern times, The Internet! Still, many people lose their jobs, money and property just because some preacher on TV says that the world is going to end, and guess what? You're not going to need money where you're going (well, neither will you sir! So, stop collecting!). And he continues on by setting out a date! Well, news flash people! If you hear that "date" part—do not pay him any attention! Because it is bound to be a lie!

In the good book, the apostle Matthew warns: "But about that day or hour no one knows, not even the angels in heaven, nor the Son, but only the Father. As it was in the days of Noah, so it will be at the coming of the Son of Man." Keep those words in mind.

Conclusion

Thank you again for downloading this book!
If you enjoyed this book, please take the time to share your thoughts and post a review on Amazon. It'd be greatly appreciated!

Thank you and good luck! ;P

www.ingramcontent.com/pod-product-compliance
Lightning Source LLC
Chambersburg PA
CBHW070126230526
45472CB00004B/1440